Kick-Ass Cannabis and Veggies:

Organic Gardening Soils, Teas, and Tips for Growing Marijuana and Nutrient-Rich Vegetables

Kip Zonderkop

Illustrations and cover design by Adriprints.

Editor: Richard Johnson

ISBN 978-1-937862-92-3
Library of Congress Control Number 2015903188

AUTHOR'S DISCLAIMER: Although I am sympathetic to the
unpleasant situations in which you may find yourself
when following this book's advice, I bear no responsibility
or liability for said unpleasantries. Sorry.

This book was published by BookCrafters,
Parker, Colorado.
http://bookcrafters.net
bookcrafters@comcast.net

This book may be ordered from
www.bookcrafters.net and other online bookstores.

CONTENTS

INTRODUCTION

What makes kick-ass Cannabis and vegetables? A mineralized soil, alive with beneficial microbes, worms, and insects. This type of soil promotes thriving indoor and outdoor gardens and produces highly nutritional fruits and vegetables. Better yet, it also produces super-resinous Cannabis buds that provide pure, unadulterated highs.

A living soil is one of the many techniques that allow Nature to grow your garden instead of commercial fertilizers.

When I started gardening, I knew I wanted to grow organically, but I had to wade through many "flavors" of organic methods, some of which are based on dubious principles. Also, many sources of great information are poorly organized, or they assume that the reader has significant gardening experience.

This book explains organic gardening techniques that are based on soil science and permaculture principles, and it employs an organized and easy-to-understand format that is accessible to even an absolute neophyte. You are presented with the fundamental techniques for starting and maintaining an organic garden, particularly indoor and container gardens, but there's plenty here for outdoor gardeners too.

The information presented in this book is not intended to be encyclopedic as it provides, in two hours or less, the absolute essential information needed to go from seed to harvest.

Do you want to learn about gardening techniques for a particular plant (such as cloning, breeding, sexing, re-vegging, pruning, topping, or training Cannabis)? Use this book as a supplement to other guides that cover those topics. However, when you want to learn established organic gardening practices and the science behind them, this IS the book you are looking for.

For those who want to get straight to it, skip to Chapter 6, *Organic Growing in Eight Steps*. Chapter 6 is a step-by-step guide for implementing the techniques presented in the first three chapters: *Soil, Garden Pots*, and *Water, Teas, and Foliar Sprays*. In addition, Chapter 4 has Worms and Chapter 5 covers *Cannabis Considerations*.

In more detail, Chapter 1, *Soil*, explains how to create and maintain a living soil and why high-quality compost is a soil's most important ingredient. For example, a mineralized soil that includes high-quality compost is viable for consecutive crop harvests, even when your garden is confined to a garden pot. Thus, the up-front work of mixing your own soil and obtaining quality compost can reward you with years of happy crops.

Chapter 2, *Garden Pots*, instructs you on which garden pots grow large, robust plants. This chapter also explains how you can mimic Nature, even with an indoor garden, by using techniques such as planting in a living mulch.

Chapter 3, *Water, Teas, and Foliar Sprays*, explains how water can be supplemented with do-it-yourself (DIY) enzyme teas, compost teas, and botanical teas made from kelp, alfalfa, or neem. These teas replace the conventional fertilizers and pesticides that are hostile to living soils.

Chapter 4, *Compost Worms*, is a quick overview on how to make and maintain a worm bin that produces high-quality vermicompost.

Chapter 5, *Cannabis Considerations*, focuses on how to discreetly grow and enjoy quality medicinal Cannabis and the basic techniques of Cannabis cultivation by walking you through the whole process, from seed to harvest.

Further Reading and Resources provides a set of references for learning more about the topics presented in this book, and resources for obtaining the soil ingredients and gardening products mentioned in this book. You can also find this section on my website, kickassorganics.com.

With the introductions and disclaimers now out of the way, let's get started.

CHAPTER 1:
SOIL

- The key to successful organic gardening is promoting soil life and providing sufficient soil nutrients
- Keep the soil at a constant, healthy moisture, which supports the complex soil life needed for robust plants

To the Rhizosphere

Soil is important because that's where the life is. Living soils provide the necessary ingredients, structure, and conditions to enhance a key component to plant life—the rhizosphere.[1] The rhizosphere is where root secretions and microbes interact by literally feeding off of each other.[2]

Roughly speaking, the root secretions feed microbes that in turn convert a soil's minerals and nutrients into forms that the roots can absorb (that is, minerals and nutrients become bioavailable). In fact, some rhizospheric microbes are in sync with their host plant, providing more nitrogen during the plant's nitrogen-hungry growth stages.[3]

Conventional fertilizers bypass and often impair the rhizosphere by directly feeding the roots with chemicals that are hostile to soil microbes.[4] By contrast, the rhizosphere is a central focus of the gardening style described in this book. In short, we are nurturing the soil's microbes for optimal rhizospheric activity. This is why the ingredients of the soil mix (provided in Chapter 6) establish and maintain microbe life for optimizing rhizospheric health. Of these ingredients, high-quality compost is essential for introducing microbial life to your soil mix.

Compost

High-quality vermicompost (also called worm compost, worm castings, or worm poo) and mature composted waste are "living" because both are full of beneficial microbe, insect, and worm life. Dr. Elaine Ingham notes that compost with a diverse set of beneficial life forms has the necessary biology to provide the requisite enzymes and hormones for a kick-ass garden:

> What makes enzymes, hormones, and plant-growth-promoting materials? The bacteria, fungi, protozoa, nematodes, and microarthropods [found in compost]. So, really, what you want to be adding [to your soil] is the biology, because they will make more of the enzyme you want. Or the hormone. Make certain that the compost contains the right set of bacteria, fungi, protozoa and nematodes so the process you want will occur. If you buy really good compost, the microarthropods [which are small crawling insects] will be present too.[5]

When you see compost worms and tiny insects crawling around in compost, don't fret. It's usually a good sign. These little guys live in and on the soil,[6] so you don't need to worry about them crawling up your larger plants.

Just be sure the little crawlies in your compost aren't dark or white-colored winged creatures, which may, in fact, be fungus gnats or white flies. They are pests that are hard to get rid of and will sap the life from your garden.

Fungus Gnat

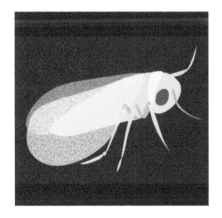 

White Flies

Chapter 3 explains how to prevent or treat such pests with DIY botanical teas made from neem oil, coffee, garlic, and oregano.

Pests aside, high-quality compost has the necessary biology for obtaining peak soil nutrient bioavailability, which is the soil–plant system's capacity to supply and absorb nutrients.[7] Since we are relying on the soil and soil life to provide the nutrients for optimum growth, compost quality will largely dictate an organic garden's health and yield.

Unfortunately, the only sure way to know the diversity and quality of your compost is to send it to a lab for analysis. There are however telltale indicators of good compost: tan to dark brown in color, coffee-ground texture, damp(ish), and an earthy smell. As noted before, top-notch compost will have microarthropods and compost worms.

Vermicompost and compost can be bought locally in most places. Just search the Internet, including Craigslist, with terms like "compost" or "organic compost". Be wary of bagged compost that lists peat moss as a main ingredient—this is not compost!

High-quality bagged compost exists, but such compost tends to be regionally distributed, so ask around and inspect the compost before you buy it.

Although compost introduces soil life into your garden, this is not enough. The soil life must be maintained with a healthy and stable soil moisture level.

The Importance of Being Moist

Although soil life can recover following drought, it will be severely inhibited during drought conditions.[8] This is why commercial worm castings and other composts that are dry or sterilized have no value to your soil mix—they're mostly dead.

Maintaining soil life requires a stable soil moisture level, which is one part of optimum soil health. Indeed, with a living soil, a dry top layer will immediately spring to life after watering—tiny insects (microarthropods) suddenly appear out of nowhere, crawling with a purpose. You can even hear a faint crackling of activity.

Don't extend this principle too far by overwatering, which might kill your plants. Think of the difference between a completely soaked cloth and a cloth that only produces a few drops if wrung. You want wrung-cloth-level wetness.[9]

When and How to Water

The optimum watering amount and frequency depend on how fast the plants drink, the temperature and humidity levels, the soil composition, and the size of the soil container. For example, soil with a higher percentage of peat moss will retain water longer than soil with a lower percentage of peat moss, assuming that both soils are in the same environment.

There are several ways to check when you need to water. An old standby is sticking a finger up to the second knuckle to check if the soil is dry. For larger containers, soil moisture meters are handy for monitoring soil moisture deep below the surface. Wilted leaves are an extreme indication, and should be avoided because this unnecessarily stresses both the plant and soil microbes.

Uniform soil moisture requires gentle watering until runoff water starts dripping out of the bottom of the container. Pour out runoff water that remains in a pot saucer an hour or two after watering.

An effective technique for full soil saturation is watering in two stages: gently water until you start to see runoff water at the bottom of the container, and then water again an hour or two later, also until runoff water appears.[10] The second-stage watering can include one of the teas mentioned in Chapter 3, because the second watering is more likely to evenly penetrate the moistened soil.

If you observe droopy leaves after watering, you may have overwatered, but the more likely culprit is poor drainage (that is, the soil needs more aeration from coarse materials such as lava rocks, etc.).

Self-Watering Gardens

There are many DIY and store-bought self-watering designs, which will provide a steady moisture level. One example is a soil moisture sensor, such as a Blumat sensor.[11] Blumat sensors include ceramic spikes that are inserted into the soil and act as a valve for tubing that is connected to the spikes and a water reservoir. The ceramic spikes expand and contract in relation to soil moisture levels. They contract under somewhat dry conditions, allowing water to flow out of the tubing and onto the soil, and then expand due to the increase in soil moisture, which stops the water flow.

Note that it is possible to incorrectly calibrate the sensor such that the valve stays open and allows the entire reservoir to drain into the soil. Although this is not a frequent occurrence, keep this worst-case scenario in mind when selecting a reservoir size. Larger soil containers (for example, garden pots of 5 gallons or more), require the larger "Tropf" Blumat sensors over the smaller "Junior" sensors.

To take the guess work out of how much water is needed, use a Blumat digital moisture meter, which measures your soil's moisture level in millibars (mb). 120 to 150 mb is considered ideal for Cannabis in the vegetative stage and 150 to 180 mb in the bloom stage, but I encourage you to use these measurements as starting points and experiment from there.

The water dispensed from these sensor systems can be supplemented with botanical and enzyme "teas", as detailed in Chapter 3.

Mineral Content

Mineral-rich soils are required to grow mineral- and vitamin-rich vegetables, which in turn provide superior nutrition for humans.[12]

You want to start a garden for healthy eating and don't know what vegetables to plant in your mineral-rich soil? Pick the most nutrient-dense veggies,[13] which are ranked below from most to least nutritious based on 17 critical nutrients:

1. Watercress
2. Chinese Cabbage
3. Chard*
4. Beet greens
5. Spinach*
6. Chicory
7. Leaf lettuce*
8. Parsley
9. Romaine lettuce*
10. Collard, turnip & mustard greens
11. Endive
12. Chive
13. Kale*
14. Dandelion greens
15. Red pepper
16. Arugula
17. Broccoli
18. Pumpkin
19. Brussels sprouts
20. Scallion
21. Kohlrabi
22. Cauliflower
23. Cabbage
24. Carrot*
25. Tomato*
26. Radish*

*indicates popular choice for indoor gardens

Why are minerals important? Minerals feed enzymes that are required for the normal functioning of many biochemical processes in the human body,[14] microbes, and plants. In fact, mineral deficiencies have well-documented symptoms in humans and plants (for example, various leaf discolorations in plants).

A balanced soil contains important minerals to maximize productivity, yield, and plant immunity. Living and balanced soils ensure that the minerals are bioavailable for root absorption as needed by the plant.

The mineral mix provided in Chapter 6 is quite wide-ranging and open. Feel free to create a diverse mineral mix — it can only help, especially for multiple-crop soils (soils intended to grow several plants, consecutively, in the same soil for multiple harvests).

To summarize the past few pages, an ideal soil has a sufficient level of nutrients and a diverse set of soil life to convert the soil nutrients into bioavailable root food.

Soil Mixing

The soil mix provided in Chapter 6 has two main components: a *base mix* (compost, sphagnum peat moss, lava rocks) and *soil amendments* (kelp, neem, etc.). Before mixing the base mix, unpack and lightly water your peat moss, preferably using water and a wetting agent such as aloe vera gel. This helps with accurately measuring the amount of peat moss you're putting in your soil mix as well as conditioning it to absorb water.

The base mix can be combined by dumping the ingredients on a tarp and shoveling them together, as shown below.

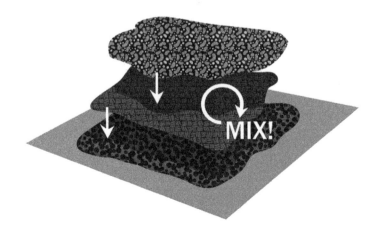

Then, mix the soil amendments together and sprinkle the mixed soil amendments on top of the base mix before shoveling everything together. This helps evenly distribute the soil amendments throughout the base mix and is an important step to creating a balanced soil.

Caution: When mixing soil, please use at least some sort of respirator or air filter and be in a well-ventilated area. Compost and soil amendments are great ingredients for soil, but not necessarily for our lungs. For example, peat moss alone can cause lung inflammation. So get a respirator when buying soil ingredients at a feed store or nursery.

When soil mixing is too much trouble, companies like BuildASoil and KIS Organics offer premixed soils that are similar to the soil mix recipe in Chapter 6.[15]

After mixing, the soil should be inoculated with an *actively aerated compost tea* (AACT). AACTs are discussed in Chapter 3, and an example is listed in Step 2 of Chapter 6.

After you've mixed the soil and watered it with an AACT, allow it to compost, or "cook", in an open container, which can be the final container for your plant (for example, a garden pot).

Air exchange is important for supplying oxygen during this process. I recommend using an open container and mixing up the soil once every week or so during the first few weeks. To further establish soil life and structure, throw in some compost worms or plant a *cover crop*, such as clover, a couple of weeks after mixing the soil. Cover crops are discussed in the next chapter, *Garden Pots*.

During the composting period, the soil pH level slowly rises to around 6.5, which is an indicator that the soil is ready to grow Cannabis, chili, and tomato plants. In fact, the Chapter 6 soil mix works well for most vegetables, because the soil's pH level generally settles between 6 and 7, which is the pH range in which most soil nutrients are plant-available. Soil pH levels can be monitored with a soil pH meter.

Of course, some plants prefer more acidic soils (berries), and some prefer less acidic soils (lilacs). Similarly, the soil mix may be a bit strong for nutrient-sensitive plants, so go light on the soil amendments (kelp, neem, etc.) for such plants. For example, I planted an indoor palm tree in a strong soil mix that had already grown two Cannabis plants (together) and a chili plant after that. Still, some of the palm tree's leaves twisted and turned and a few tips turned brown, which are classic signs of over-fertilization.[16] That said, the palm tree exploded with growth, and within a few weeks it looked the healthiest I had ever seen it. So soil strength can be a difficult balancing act.

Seedling Soil

The base mix in Chapter 6 can be used immediately after mixing and is fine for germinating seeds and young plants. In fact, high-quality compost has a high germination rate, and special germination techniques, such as soaking seeds in paper towels, are not necessary.

The complete mix (base mix plus amendments) must compost or "cook" for at least 3 weeks, but preferably longer, especially for seeds and young or sensitive plants. A three-week composting period is the bare minimum before planting larger plants, but planting cover crop seeds at that point is fine.

Multiple-Crop Soil and Soil Remixing

Depending on the quantity and quality of the soil, it can support consecutive crop harvests with minimum effort.

For example, after harvesting or chopping down the first plant grown in the soil,[17] plant another seed or dig a hole in the soil for a second plant (a *transplant*) that is already growing in another pot.[18] Take the transplant out of its first pot, plant the transplant in the hole, and cover the transplant's root ball with soil and/or worm castings. The worm castings and other topdressing layers, as explained in the next chapter, provide additional nutrients for the transplant.

These practices are known as *no-till gardening*. Tilling the soil greatly disturbs, if not destroys, soil life. The no-till approach instead replants seeds or plants into the soil after harvest and enriches the soil with topdressing layers.

Eventually there comes a point when topdressing is not sufficient and the soil needs more of the original ingredients, as detailed in the optional eighth step in Chapter 6. Unfortunately, this means it's time to restart the soil-building process by breaking up the old root balls, mixing in the new soil ingredients, and composting the soil for 3 weeks or longer. Similarly, the old soil can be combined with newly mixed soil.

The composting period can be avoided by lightly amending the soil after a few harvests. This method is also described in Step 8 of Chapter 6.

Endnotes:

1 http://en.wikipedia.org/wiki/Rhizosphere.

2 http://www.google.com/patents/US7595061 ("Soil contains a diversity of life forms which can interact with plants, such life forms including bacteria, fungi and nematodes. These biological forms are particularly abundant in the rhizosphere, the area of soil that surrounds and is influenced by the plant roots. Rhizobacteria are those bacteria which are adapted to the rhizosphere. There is a complex interaction among the various life forms in the soil, where some are antagonistic and others are mutually beneficial. Similarly complex is the interaction between the plants and the soil life forms, which can helpful to the plant in some instances, and harmful in others.")

3 For example, nitrogen-fixing bacteria found in the rhizosphere of the rice plant fix more nitrogen during rice growth stages, which is when rice requires more nitrogen. See Sims GK, Dunigan EP (1984). *Diurnal and seasonal variations in nitrogenase activity (C2H2 reduction) of rice roots.* Soil Biology and Biochemistry 16(1): 15–18.

4 Jeff Lowenfels and Wayne Lewis, Teaming with Microbes: The Organic Gardener's Guide to the Soil Food Web (Timber Press 2010), pg. 26 ("Chemical fertilizers, pesticides, insecticides, and fungicides affect the soil food web, toxic to some members, warding off others, and changing the environment."; "When chemically fed, plants bypass the microbial-assisted method of obtaining nutrients, and microbial populations adjust accordingly.").

5 http://theearthproject.org/id64.html

6 http://www.pacifichorticulture.org/articles/soil-microarthropods.

7   Hormoz BassiriRad, ed., *Nutrient Acquisition by Plants: An Ecological Perspective* (Springer 2005), pg. 1 ("The soil-plant system's capacity to supply/absorb nutrients is termed *soil nutrient bioavailability*, and is the ability of the soil-plant system to supply essential plant nutrients to a target plant, or plant association, during a specific period of time as a result of the processes controlling (1) the release of nutrients from their solid phase in the soil to their solution phase; (2) the movement of nutrients through the soil solution to the plant root-mycorrhizae; and (3) the absorption of nutrients by the plant root-mycorrhizal system.") [citation omitted].

8   http://www.nature.com/nature/journal/v501/n7468_supp/full/501S18a.html ("The drying of the soil can disrupt the diversity of the microbiome and cause microbial biomass to fall by two-thirds or more. Soil microorganisms will eventually return to their pre-stress population levels, but any crops planted in the meantime are vulnerable.").

9   http://www.ext.colostate.edu/mg/gardennotes/212.html#encourage ("Water effectively. Soil organisms require an environment that is damp (like a wrung out sponge) but not soggy, between 50–90°F. Soil organism activity may be reduced due to dry soil conditions that are common in the fall and winter. Avoid over-irrigation because water-logged soils will be harmful to beneficial soil organisms.").

10 The Rev, *True Living Organics: The Ultimate Guide to Growing All-Natural Marijuana Indoors* (Green Candy Press 2012), pg. 124 ("You water with plain water, lightly, until you see some water run out of the bottom of the container. Then wait for at least 1 hour (I often wait a couple of hours) then water again until you see drainage from the bottom.").

11 This link is a Google image search of "Blumat sensors."

12 For more information on this subject, I refer you to Steve Solomon's book *The Intelligent Gardener: Growing Nutrient Dense Food.*

13 http://www.cdc.gov/pcd/issues/2014/13_0390.htm

14 Sally Fallon and Mary Enig, *Nourishing Traditions: The Cookbook that Challenges Politically Correct Nutrition and the Diet Dictocrats* (Newtrends Publishing, Inc. 2003), pg. 46.

15 Sally Fallon and Mary Enig, Nourishing Traditions: The Cookbook that Challenges Politically Correct Nutrition and the Diet Dictocrats (Newtrends Publishing, Inc. 2003), pg. 46.

16 Adding some sand would have helped, too.

17 The previous plant's root ball can remain, or rather, what's left of it after you dig a hole for the next plant. Roots can rot, but good soils break down the roots well before rot would set in. Enzymes play a big part in this, so perhaps water with an enzyme tea, as described later, to help speed this process along. That said, I've never noticed a problem, even when using just plain ol' water.

18 The next plant could be a Cannabis transplant that you'll soon switch over from vegetative to flower stage.

CHAPTER 2:
GARDEN POTS

- Larger containers grow larger plants
- 5-gallon containers are considered the minimal size for no-till gardening
- Use garden pot dollies to move larger garden pots with ease
- Mimic Nature by including a clover cover crop and compost worms within a garden pot
- The terms *pots* and *containers* will be used interchangeably in this book

Choosing a Container

At this stage, your soil is fully composted and you're ready to fill a garden pot. You can also compost the just-mixed soil in your garden pot.

GeoPot, Smart Pots, and Air-Pots brand containers encourage plant growth because of their breathable designs that allow oxygen exchange with roots. This type of container air-prunes the root tips which increases root branching and avoids pot-bound roots.[19] All of this leads to a better-developed root system and more robust plant growth compared to plastic containers.

For outdoor gardens, a simple raised bed can be made by building a 6- to 12-inch deep wooden frame that you place on the ground and fill with soil.[20]

If you're considering consecutive crops or multiple plants in the same soil and container, choose containers that are at least 5 gallons, but preferably larger. For example, two bushy Cannabis Indica plants can peacefully coexist in 15-gallon containers, or in smaller wide-dimensioned containers.

In general, container size affects how large your plant will grow—larger containers grow larger plants.[21] That said, I imagine a 20-gallon container would not provide noticeable additional growth over a 15-gallon container for most vegetables and annuals (for example, Cannabis), but feel free to experiment.

Of course, mobility becomes an issue with larger container sizes. A simple way to mobilize larger containers is to place a small dolly (a wheeled platform) underneath the containers.

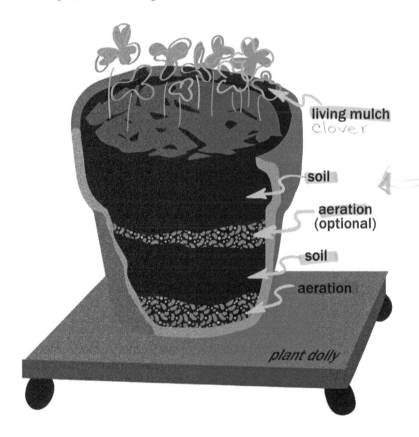

living mulch
clover

soil

aeration
(optional)

soil

aeration

plant dolly

This also allows for water drainage and airflow at the bottom of the container. Many nurseries have dollies for garden pots. You can build your own as a relatively easy DIY project.

I fill containers as shown in the illustration. First, I pour ½ to 2 inches of an aeration element, such as lava rocks, in the bottom of the container. Then, I fill the container at least halfway with my soil mix and throw in a few earthworms, which aerate the soil and provide worm casting. For larger containers (10 or more gallons), I place a ¼- to ½-inch layer of aeration material after filling the container half way.

Mulches and Topdressings

Adding mulch is typically the last step of building the container. The mulch can be (1) a cover crop (also called a living mulch), (2) a topdressing of worm castings and/or harvested beneficial plants, such as *dynamic accumulators* (further detailed in the next section), or (3) a combination of cover crops and topdressings.

For example, leaves, stems, and roots harvested from dynamic accumulators can be placed on top of the soil or buried by a top layer of soil. This allows the nutrients stored in the leaves, stems, and roots to be released in the soil. Topdressing can include both dynamic accumulators and worm castings mixed together or layered with the dynamic accumulators under a top layer of worm compost.

Pruned leaves from your vegetable or Cannabis plant also make good additions to the top layer of mulch.

Dynamic Accumulators

Dynamic accumulators are plants that store nutrients taken from the ground or air in their leaves, stems, and roots.[22] These natural fertilizers are not created equally; each has a uniquely accumulated nutrient profile, and several sources list the nutrient profiles for a variety of accumulators.[23]

Dandelion, kelp, stinging nettle, comfrey, and watercress are standout dynamic accumulators that have a broad spectrum of nutrients, among other beneficial properties. Dandelion is particularly handy because it grows just about everywhere. When possible, pick dandelions from areas free of herbicides and pesticides.

Super gardeners grow patches of dynamic accumulators, which they then harvest (cut or mow) and use as topdressing for other plants.[24]

Dynamic accumulators are used widely to make botanical teas for soil application and foliar sprays, as further explained in Chapter 3.

C(l)over Crop

A cover crop can be planted after placing the soil in a container, assuming that a new soil mix has composted for at least 2 weeks. White clover, particularly white Dutch clover, is one of Nature's best nitrogen fixers. *Fixing*, in this context, is the process of converting atmospheric nitrogen into root-absorbable nitrogen. Cover crops also promote stronger soil structure and soil life.[25]

One clever gardening technique includes burying clover, which stores nitrogen in its leaves and roots along with fixing nitrogen in the soil, so that the nutrients stored in the clover are released in the soil and available for the main plants of your garden.

For example, plant clover in the soil just after placing the mixed, but not fully composted, soil in the growing container. When the soil is ready for larger plants (for example, tomatoes), the larger plant is transplanted on top of or in the middle of the clover, and the clover is buried with a layer of soil or worm castings. The buried clover provides nitrogen, among other things, for a larger plant. That's why buried cover crops are called green manure.[26] An additional cover crop can then be planted after transplanting.

Another clever technique includes planting a variety of cover crop plants, such as white clover and buffalo grass. These companion plants grow more robustly together than by themselves.[27] This synergy benefits both the cover crop plants and the larger plants near the cover crops.[28]

One type of cover crop, the herbal ley, includes flowering herbs that attract predatory insects, which eat orchard pests such as other insects, snails, and slugs.[29] Similarly, beneficial weeds[30] attract beneficial insects,[31] and trap crops attract pests away from valued plants.[32]

For container gardening, a cover crop of microclover or Dutch white clover is fine; add some barley grass if you want to get fancy. Ideally, grow and maintain your cover crop before, during, and after the life cycle of the associated fruiting plant.

Cover crops also play a canary-in-a-coal-mine role for your garden: cover crops will show signs of distress, such as wilting from drought or pests, before similar signs occur or can be recognized in a larger plant.

Pit Greenhouses: The Ultimate Indoor Containers

For ambitious gardeners, a clever permaculture "container" is a pit greenhouse.[33] By digging 6 to 8 feet into the ground, these greenhouses can grow a garden year-round using only heat from the earth and sun, even in frigid environments. A no-frills DIY guide to building one can be found here.[34] For those who really want to dig in, The Earth-Sheltered Solar Greenhouse Book is a detailed reference book on the subject.[35]

Endnotes:

19 http://depts.washington.edu/propplnt/Chapters/air-pruning.htm.

20 http://www.wikihow.com/Practice-Square-Foot-Gardening

21 http://www.sciencedaily.com/releases/2012/07/120701191636.htm ("Doubling plant pot size makes plants grow over 40% larger.").

22 http://en.wikipedia.org/wiki/Dynamic_accumulator.

23 http://soulflowerfarm.blogspot.com/2012/11/dynamic-accumulators.html; http://en.wikibooks.org/wiki/Permaculture_Design/Soil#Dynamic_Accumulators_of_Nutrients_for_Composting.

24 http://buildasoil.com/blogs/news/9745744-diy-nutrients-from-dynamic-accumulator-plants.

25 http://en.wikipedia.org/wiki/Cover_crop.

26 http://en.wikipedia.org/wiki/Green_manure.

27 Deyn GB, Quirk H, Oakley S, Ostle NJ, Bardgett RD (2012). *Increased Plant Carbon Translocation Linked to Overyielding in Grassland Species Mixtures*. PLoS ONE 7(9): e45926.

28 *Ibid*.

29 Mary Horsfall, *Creating Your Eco-Friendly Garden* (Csiro Publishing 2008), pg. 90 ("Commercial organic orchardists are increasingly using herbal leys as a cover crop between fruit trees. This consists of a mixture of flowering herbs — tansy, yarrow, alyssum, borage, parsley, lovage, dill and marigolds, for example. These species attract predatory insects.").

30 http://en.wikipedia.org/wiki/Beneficial_weed.

31 http://en.wikipedia.org/wiki/Beneficial_insect.

32 http://en.wikipedia.org/wiki/Trap_crop.

33 http://www.treehugger.com/green-architecture/build-underground-greenhouse-garden-year-round.html; http://www.inspirationgreen.com/pit-greenhouses.html

34 http://www.the-meal.net/graph/manuel_walipina_benson.pdf.

35 http://www.undergroundhousing.com/greenhouse_book.html

CHAPTER 3:
WATER, TEAS, AND FOLIAR SPRAYS

- Always use unchlorinated water
- Kelp meal, alfalfa meal, and/or neem meal soaked in aerated water for 24 hours makes a tea that can be applied to soil or leaves

At this stage, your plants are growing in your custom-made container. Congrats! Although you can grow a kick-ass garden using only water, this chapter covers the extras needed for a more robust and productive garden.

Water

One advantage of this gardening style is that the soil can accommodate both soft and hard water. The soil is quite robust, and hard water should not increase soil pH to a detrimental degree.[36] Exceptionally hard water (500 parts per million or greater of dissolved solids) and contaminated water should be treated with at least a carbon filter. Water from water softeners is not appropriate for gardens.

Fish tanks, reclaimed rain, lakes, and rivers can be great sources of water for your garden. When you are using tap water, chlorine or chloramine may be present. These chemicals suppress the good microbe life in soil and teas.

Chlorine is relatively easy to get rid of: let tap water stand in an open container for 24 hours or aerate, for example, with a fish tank air pump. Chloramine, however, is more stable than chlorine, and cannot be evaporated or bubbled away.

Nevertheless, there are several techniques for removing either chemical. A good carbon or reverse osmosis (RO) filter will remove most of either chemical when it is present in the tap water, and humic acids neutralize chlorine and chloramine when mixed with water.[37]

Humic acid comes in a bottled form (for example, BioAg Ful-Power fulvic acid) and is also present in high-quality compost. So just mix in a tablespoon of compost per gallon of water and let it sit for an hour or two before brewing a tea or watering your garden.

Some cities do not add chlorine except during flood conditions, so investigate.

In sum, water your garden with chlorine-free water, and chlorine-free water should be used for the teas and foliar sprays you create using the methods discussed in the following sections.

Botanical Teas

Kelp meal, alfalfa meal, and/or neem meal can be bubbled in water with an aquarium pump for a duration of 24 hours (or longer) to make a botanical tea. Begin with a meal-to-water ratio of 1 teaspoon per liter or 1 tablespoon per gallon. Place the meals and other botanicals in a teabag before bubbling in the water to help avoid clogging your watering cans and sprayers. The contents of the used teabag can then be fed to the worms in your worm bin. You can also strain the tea before application if the meals are free floating during the bubbling period.

Kelp, alfalfa, and neem meal can be used together in one application (for example, 3 tsp/liter); the resulting tea may need to be diluted with more water for delicate plants and foliar spray applications. Alternatively, individual teas can be applied in cycles. For example, when you apply botanical teas once a week, which is a reasonable frequency, the weekly applications can rotate through each of the three ingredients.

Alfalfa, kelp, and neem are easy to obtain and widely used by gardeners as organic fertilizers or pest control in tea form. Kelp and alfalfa are dynamic accumulators, and neem has documented properties against harmful insects, fungi, and bacteria.[38]

Teas can be applied as a soil soak (that is, watered in soil) or as a foliar spray, which feeds plants through their leaves.[39] Foliar sprays can be applied daily, depending on the plant, temperature, and humidity conditions. Some gardeners even spray Cannabis leaves up to the mid-flowering stage. However, Cannabis is sensitive to prolonged wet foliage and can contract bud rot in humid conditions or environments with low air exchange.

Note: Alfalfa may unnecessarily extend a plant's flowering stage and shouldn't be used past the early onset of flowering.

Foliar sprays need a surfactant such as aloe vera gel, powder, or juice;[40] yucca extract; or a mild soap, such as a castile soap. Surfactants lower the surface tension of water and prevent the foliar spray from beading off of the surface of the leaves before the nutrients are absorbed.

In addition to neem meal, neem oil is also a popular foliar spray. However, repeated, heavy foliar applications of neem oil on a plant in the fruiting stage may produce bitter-tasting fruits and vegetables.

Neem oil is powerful stuff, so start out conservatively in both the amounts you use and the frequency with which it is applied, in particular as a soil soak, where strong doses might interfere with the beneficial bacteria and fungi growing in your soil.

Making other botanical teas beyond the basic trinity of kelp, alfalfa, and neem is encouraged since each plant offers a unique combination of benefits. For example, fresh dandelion, stinging nettle, comfrey, and watercress can be chopped and soaked for several days to make a fermented tea[41] or bubbled for 24 to 72 hours to make a botanical tea.

Natural Insecticides

Some gardeners cycle through foliar sprays of neem oil, lavender oil, rosemary oil, and peppermint oil as insecticides and plant immunity boosters. Foliar sprays made from chili peppers, garlic, coffee grounds, oregano, or cilantro (coriander) are also effective insecticides.

Combining oil and a mild soap forms an effective mixture for spraying against small insects like aphids, mites, white flies, and fungus gnats, which are suffocated when exposed to this mixture. So be sure to coat them thoroughly.

Note: Always test any foliar spray on a single leaf before applying it to the entire plant. You want to make sure that the spray is not harmful to the plant.

Aloe and Coconut Water

Two further supplements, aloe vera gel and coconut water, have a broad spectrum of beneficial enzymes and nutrients and can be added to plain water or botanical teas, as detailed in Chapter 6.

Both are biostimulates. For example, aloe triggers a plant's natural defenses and stimulates root growth for clones, making it a natural root promoter for an organic cloning gel: mix together 1 gallon water, 1 ounce Ful-Power fulvic acid, 2 ounces aloe vera gel, and 1 teaspoon Pro-TeKt liquid supplement.[42] This cloning gel is as good as or superior to commercial cloning gels.

Aloe can be used for both foliar spray and soil soak applications, whereas coconut water is best used for soil soaks.

What Is an Actively Aerated Compost Tea (AACT)?

AACTs are rich with microbes and are made by aerating a mixture of water and compost. AACTs inoculate a just-mixed soil with beneficial microbes and can be subsequently added once a month during crop growth to increase your soil's microbe levels. While AACTs are normally applied to soils, foliar application of AACTs may result in good microorganisms outcompeting mold for the same food on the surface of your plants.

AACTs can be dominated by bacteria or fungi, depending on the selected food source and the bacteria/fungi balance of the compost[43] used in the brew. Although vegetables and annuals (for example, Cannabis) generally thrive in bacteria-dominated soils,[44] and trees and shrubs prefer fungus-dominated soils, this does not mean that an AACT should follow suit. Instead, AACTs should have both bacteria and fungi, resulting in a balanced tea and soil.

To explain, AACTs supplement the soil by providing further bacteria and fungi for the soil-to-root interface called the rhizosphere. As discussed in Chapter 1, the rhizosphere is an area where root secretions (called root exudates) feed specific bacteria and fungi.[45]

The composition of the root secretions changes according to the plant's needs, including when a plant enters the fruit or flowering stage. The plant's root secretions initially feed the soil bacteria that provide the nitrogen needed for growth during a vegetative or growth stage. When the plant transitions to the flowering stage, the secretion composition changes to feed the soil fungi that provide phosphorus and potassium, which is required during the plant's fruiting stage.[46]

These different requirements indicate that using a bacteria-dominated tea during the vegetative stage and a fungus-dominated tea during the flowering stage is the way to go, and there are some gardeners who follow this strategy. However, the exact microbe composition of soils and teas are unknown to most of us, and it is simpler to provide a balanced tea to meet all of your plant's needs.

How to Make an AACT

AACT is made by bubbling, with an aquarium pump or other aerator, a mixture of compost and a bacterial or fungal food source in nonchlorinated water. The food source assists in multiplying bacteria and growing fungi, and the bubbling provides oxygen for the microorganisms. The only food source needed to brew a balanced tea is unsulfured blackstrap molasses.[47]

AACTs should be made in a dark environment, or at least in an opaque container (for example, a pitcher) placed away from direct sunlight.

Aquarium pumps provide enough oxygen for up to 2 gallons of tea, but other designs are needed when you brew larger batches. For example, the Mini-Microbulator and a similar brew system offered by KIS Organics will brew up to 5 gallons, and both offer much larger brew systems.[48] There are also DIY designs that can be found by searching "DIY AACT brewer" on a browser.

I brew in a 5-liter watering can that has an open spigot (that is, it has no sprinkler head). After brewing is complete, I pour the tea through a strainer into another watering can. This avoids clogging the sprinkler head of the other watering can.

Foliar spray applications may need to be filter more thoroughly to avoid clogging your sprayer. Use a filter mesh no smaller than 400 microns so you avoid filtering out the beneficial bacteria and fungi in your tea.[49]

The rate of microorganism growth is dependent on the amount of food available and the ambient temperature (more food and time is needed for temperatures below 72° F).[50] However, rigorous bubbling of a ½ cup compost and 1.5 tablespoons of molasses per gallon of water for 36 hours works well when the ambient temperature is between 65° and 72° F.[51]

Brewing time can be shortened to 24 hours by pre-feeding your compost before brewing an AACT. For each ½ cup of compost, mix in ½ tablespoon of wheat bran, oatmeal, or oatmeal powder in a container. Moisten the mixture with heavily diluted blackstrap molasses (water with a drop of molasses) and cover the mixture with a cloth or paper towel for 24 hours.[52]

When fungal growth is your goal, there are a couple methods you can use to increase fugal levels of your AACT. One method is to add a commercial mycorrhiza[53] product before watering the plants with the tea (that is, adding the mycorrhiza after the brewing process is complete).

Another method is to pre-grow fungi, as explained in Teaming with Microbes, by mixing oatmeal with compost before brewing the compost in water. Mix the oatmeal at a rate of 3 to 4 tablespoons for each cup of moist compost in a container and let sit for 3 days at a temperature of around 80° F, which should result in compost covered in "long, white, fluffy strands" ready to be brewed in an AACT.[54]

Fungal growth also occurs when oatmeal is placed on top of or under a thin layer of the compost in a worm bin. I've noticed that scraps in healthy worm bins tend to get mold more quickly than in very acidic or otherwise imbalanced bins (for example, bins with too much food scraps or are too wet or dry), likely due to lower fungal levels.

This brings up an important point: wheat bran and oatmeal can only enhance the fungi already present in your compost; they cannot provide soil fungi.

Enzymes: Bringing It All Together

Enzymes are equally essential for humans, plants, and microbes. Sally Fallon and Mary G. Enig explain enzymes as "complex proteins that act as catalysts in almost every biochemical process that takes place in the body. Their activity depends on the presence of adequate vitamins and minerals. Many enzymes incorporate a single molecule of a trace mineral—such as copper, iron or zinc —without which the enzyme cannot function."[55]

For gardening, we are interested in enzymes that accelerate organic matter decay or stimulate soil microbes, which also produce enzymes that convert soil components into a form that plant roots can absorb.

Compost with a diverse set of life (bacteria, fungi, small insects, nematodes, earthworms) will be rich in enzymatic activity.[56] In addition, there are several commercial sources of enzymes (soil additives and similar products with names that end with "zyme").

A DIY method for increasing enzyme levels in soil has been devised by a clever gardener named "ClackamasCootz"; the method has been posted on various gardening Internet forums and is described in detail in Chapter 6. Basically, barley seeds are sprouted and then blended in water, with the resulting blend added to water for watering soil.[57] The young sprouts are rich in enzymes, making this is a cheap and easy way to enrich your soil.

Enzymes teas can be given a boost with fulvic and humic acids, which enhance enzymatic activity.[58] Such acids are present in high-quality compost and are also available in concentrated form in products such as BioAg's Ful-Power fulvic acid and TM-7 humic acid. Both products are highly recommended add-ons to enzyme teas or botanical teas, and only require weekly applications.

Endnotes:

36 http://www.finegardening.com/how-to/qa/irrigating-hard-water.aspx ("Hardness is related to the content of calcium carbonate and magnesium carbonate dissolved in the water . . . When hard water is used for irrigation, it's the same as adding a small amount of lime every time you water. Over time, this continual addition of lime will increase the pH of the growing medium.") (dead link).

37 http://www.natureswayresources.com/DocsPdfs/chloramine.pdf.

38 National Research Council, *Neem: A Tree for Solving Global Problems* (National Academy Press 1992), pg. 3 ("[Neem's] chemical weapons are extraordinary, however. In tests over the last decade, entomologist have found that neem materials can affect more than 200 insect species as well as some mites, nematodes, fungi, bacteria, and even a few viruses.").

39 http://en.wikipedia.org/wiki/Foliar_feeding.

40 Aloe vera gel can be obtained straight from the plant or commercially sourced in gel, juice, or powder form.

41 http://www.frenchgardening.com/tech.html?pid=3164873867231346 ("The duration of fermentation can range from a few days to a couple of weeks. When the mixture stops bubbling when you stir or otherwise move the contents, fermentation is complete. Check your brew daily.")

42 Concoction credit goes to ClackamasCootz. Pruned Cannabis branches (cuttings) near the bottom of the plant sprout roots when done correctly. Place the cuttings in the cloning gel and soak the cloning plugs in the gel for a few hours before placing the cuttings into the plugs. A detailed discussion on cloning can be found on the Internet or a general Cannabis guide.

43 Worm compost is bacterial-dominated, but will also have fungi.

44 Lowenfels and Lewis, *Teaming with Microbes*, pg. 26 ("In general, perennials, trees, and shrubs prefer fungally dominated soils, while annuals, grasses, and vegetables prefer soils dominated by bacteria.").

45 *Ibid.*, pp. 20–21 ("Amazingly, [the] presence [of root exudates] wakes up, attracts, and grows specific beneficial bacteria and fungi living in the soil that subsist on these exudates and the cellular material sloughed off as the plant's root tips grow. All this secretion of exudates and sloughing-off of cells takes place in the rhizosphere, a zone immediately around the roots, extending out about a tenth of an inch, or a couple of millimeters.").

46 http://microbeorganics.com ("When the plant receives its signal from the upper world, above the soil, that it is time to switch gears and produce flowers and/or fruit, its nutrient requirement changes. . . . [S]tudies indicate that the plant then increases the uptake of the ammonia (N) (bioavailable nitrogen) and reduces or stops excreting the carbon which [previously fed] the bacteria/archaea. This effectively starves the bacteria/archaea which will react by dying or becoming dormant. . . . The mycorrhizal fungi . . . [are] then triggered into increased growth and production. Studies have indicated that the transference of bioavailable phosphorus and potassium to the roots occur mainly as a function of arbuscular mycorrhizal fungal hyphae in symbiotic relationship with the roots of the plant.").

47 *Ibid.* ("Black strap molasses (BSM) is a complex sugar/carbohydrate and feeds bacteria/archaea and fungi equally well.").

48 *Ibid.* and http://www.kisorganics.com.

49 *Ibid.*

50 http://edibleschoolyard.org/node/12188 ("The amount of food varies a bit with the outside air temperature. The higher the temperature above 72-degrees F, the less food you should add. The further below 72-degrees F, the more food you should add. At 100-degrees F, add no additional foods at all. At 50-degrees F, double the amount of food. In general, 0.1% food should be added, given the adjustment for temperature. So, 4 gallons of brew would get about 1 tablespoon total of foods.").

51 http://microbeorganics.com.

52 *Ibid.* ("To spill a small secret, I've been pre-feeding or pre-activating [vermi]compost which is not so fresh by mixing in a small amount of wheat bran (livestock store or bulk foods department grocery store) and moistening with very diluted black strap molasses, loosely covered with cloth or paper towel 24 hours ahead of brew. (approximate ratios, wheat bran 1:30 [vermi]compost & BSM 1:300 water).").

53 http://en.wikipedia.org/wiki/Mycorrhiza.

54 Lowenfels and Lewis, *Teaming with Microbes*, pg. 155.

55 Fallon and Enig, *Nourishing Traditions*, pg. 46.

56 http://theearthproject.org/id64.html.

57 http://buildasoil.com/blogs/news/12607517-using-b-a-s-barley-for-enzyme-tea-tutorial.

58 William R. Jackson, *Humic, Fulvic and Microbial Balance: Organic Soil Conditioning* (Jackson Research Center 1993).

CHAPTER 4:
COMPOST WORMS

- Worms can thrive in plastic storage containers and plant pots
- A simple mix of 25% straw (or aeration element) and 75% composted manure can feed worms for weeks
- Diverse mineral inputs increase compost quality
- Put worms in your garden pots

Worm bins are one the easiest, fastest, and cheapest ways to obtain high-quality compost. Compost worms are easily obtained (they can be locally sourced or shipped) and they generate compost much faster than most home compost heaps, so they are winners all the way.

Worm Bins

The worms need a home containing the following:

(1) a food source such as composted manure or food scraps, and
(2) a bedding.

Shredded newspaper is a common bedding material, but I instead use composted manure that is mixed with straw or another aeration element, such as lava rocks, at a 3 to 1 ratio (that is, 75% manure and 25% straw or lava rocks). An advantage with lava rocks is that when adding worm castings to your soil mix, the harvested castings are already mixed with an aeration element. I don't use newspaper strips because I am too lazy to cut up paper, and with this bedding there are always some undigested strips in the harvested vermicompost.

The worm bin holding the worms and bedding can be made from almost any container that allows for air flow, including plastic storage containers and garden pots. There are also stacked tray worm bin designs that separate the compost layers. However, a simple bin can achieve "worm/worm casting separation" just as well as the tray systems.

For example, in *The Best Place for Garbage*,[59] Sandra Wiese points out how worms can easily be moved from one side of the bin to another by placing food only on one side of the bin.[60] This is the same idea behind the trays: feeding the upper trays attracts the worms from the lower trays in order to harvest the "worm-free" castings from the lower trays. So don't feel compelled to buy a tray system when a simple DIY container will also work.

Although composted manure is both a food source and bedding, the worms will still gravitate to wherever the kitchen scraps are placed, even in manure beddings.

Worm Bin Health

The two most common mistakes are over-watering and over-feeding the worm bin. Like soil, worm bins should be moist, but not wet like mud. The wrung-cloth analogy used in Chapter 1 is equally applicable here: moist, but not soaking.

For feeding, try feeding the worms two to three times their cumulative weight in food once a week. If there is still a significant amount of food left, use less food. The goal is to have a small amount remaining by the next feeding.

Monitoring how much food is left between feedings is easier by placing food scraps in a consistent place within the worm bin.

Worm Food

Aged or composted manure makes for a very simple worm bin because the manure is both a food source and a bedding material. For example, in a bin of mixed composted manure and straw, worms can live for weeks, if not months.

Don't use fresh manures, which may contain deworming medications, among lots of other nasties.

Beyond manure, almost anything that was once living can be thrown in the worm bin for worm food. Many guides warn against onions and citrus, but worms are not picky eaters,[61] so feel free to chuck in most things.[62] Feed your worms citrus fruit and onions sparingly, and avoid meat and very salty or oily foods.[63]

Fruit rinds, vegetables, and coffee grounds are suitable worm foods, but they will need additional amendments to generate kick-ass vermicompost.

Amendments

Amendments such as neem meal, kelp meal, alfalfa meal, crab shell meal, oyster shell flower, glacial rock dust, basalt, bentonite, gypsum, and other rock dusts can be added to your worm bin to enhance worm castings as a fertilizer.

Do these ingredients look familiar? They are the same ingredients required for making kick-ass soil. These ingredients provide nutrients and homes for beneficial microbes; and some also buffer against changes in soil pH levels (not neem or kelp). As a bonus - worms love neem, which is shown to stimulate both worm growth and reproduction.[64]

While I do not assert which specific amounts of amendment are "the best", I do invite you to experiment with ratios a bit less than those listed for use in the soil amendment recipe listed in Chapter 6 (for example, only a ½ cup or less of rock dusts per cubic feet of worm bin material per application).

The spent alfalfa, kelp, and other botanical leftovers removed after brewing botanical teas, as described in the previous chapter, are perfect additions to your worm bin. For example, vermicompost may be low in magnesium,[65] which can be compensated for by adding kelp.

Harvesting Worm Compost

There are lots of ways to separate the worms from the worm compost for garden use. One way is to place food only in a particular section: the worms will migrate to the food, thus leaving a relatively worm-free section in the bin. This is one of the easiest ways to harvest. Worm eggs will still remain in the worm-free section, but there is no advantage to removing the worm eggs before using the harvested vermicompost. Worms are great for soil, so leave the eggs in the harvested compost, they will hatch and populate the soil mix with natural soil aerators and fertilizer producers.

Other harvesting methods take advantage of worm sensitivity to light. For example, dump the bin contents on a tarp under a bright light and separate the bin contents into tall rows. As you remove the tops of the vermicompost rows, the worms will dive down deeper in the row to avoid the light. Keep combining the material to provide a tall pile with deep escape for the worms, allowing you to remove more worm-free vermicompost from the top.

Endnotes:

59 I recommend this book for those new to worm farming.
60 Sandra Wiese, *The Best Place for Garbage* (WiR Press 2011), pg. 198.
61 Technically, the microbes break down the food for the worms to eat.
62 Wiese, *The Best Place for Garbage* at Chapter 6 ("Generally, any food from your kitchen that you are not going to eat can go to the worms, oily and salty things excepted.").
63 *Ibid.*
64 Gajalakshmi and Abbasi, Bioresource Technology, *Neem leaves as a source of fertilizer-cum-pesticide vermicompost;* 92(3): 291–6 (May 2004).
65 https://www.bae.ncsu.edu/topic/vermicomposting/vermiculture/castings.html ("The nutrient content was much higher in the vermicomposts for most elements except magnesium [compared to commercial compost]. . . . The researchers noted that many of the nutrients in waste materials (including nitrogen, potassium, phosphorus, calcium and magnesium), when processed by earthworms, are changed into forms more readily taken up by plants.").

CHAPTER 5:
CANNABIS CONSIDERATIONS

- Grow tents, air exchange fans, and carbon filters are commonly used for discreet indoor gardens
- LED grow lights last longer, emit less heat, and use less electricity than conventional grow bulbs
- Choose Cannabis seeds with good genetics

This chapter provides information for the true beginner in the art of Cannabis gardening. It is not meant to replace a Cannabis gardening guide, but rather to efficiently inform the novice of basic skills. For more detailed information, refer to Greg Green's *The Cannabis Grow Bible*, Ed Rosenthal's *Marijuana Grower's Handbook*, and other quality guides.

The Internet contains terabytes of data covering this topic and can provide many interesting ideas. For example, http://www.marijuanagrowershq.com/ has a wealth of free information on many Cannabis gardening techniques. Streaming online videos on how to plant seeds, prune, transplant, etc., can also expand your gardening knowledge by presenting these topics visually. However, after reading this chapter, you will have learned enough of the basics to grow a kick-ass Cannabis plant.

Discreet Gardening

Growing Cannabis comes with unique challenges. Even in the few locations that it can be grown legally, the need to garden discreetly remains crucial.

Grow tents provide an easy way to create a dedicated growing area indoors. They are made of light-proof fabric covering an inner frame and have zippered doors and various openings to accommodate air circulation, electrical cords, and access to plants.

Grow tents are fitted with a circulation fan designed to move hot humid air out of the growing area. This air movement produces negative air pressure inside the tent, but if the tent walls bow in severely when the fan is on, the fan is being unnecessarily strained, which may shorten its life. This problem can be fixed by opening circulation flaps until the optimal air flow is achieved.

Negative pressure ensures that odors within the tent cannot leak out. Instead they are pulled out through the fan and exhausted to a discreet location or through a carbon filter. You can also avoid plant odors by selecting less pungent strains of Cannabis.

Humidity is another condition you need to control within the tent. Exhausting humid air through your carbon filter can cause several issues that reduce filter longevity and efficiency. When odor control is a necessity and humidity is high, you will likely need to use a de-humidifier to control the relative humidity (RH) inside your tent. Also, if you just vent humid air into the room that houses the grow tent, the humidity will damage the room if you don't regularly allow fresh air in.

Air circulation is also important because carbon dioxide (CO_2) is required by the plant to grow. The circulation removes the oxygen (O_2) generated by photosynthesis and replaces the depleted CO_2. Just by breathing in your tent you will increase the available CO_2 for your plants. Carbon dioxide supplementation is employed by many production grows to increase yield, but it is an advanced technique that is potentially hazardous and beyond the scope of this guide.

LED grow lights are another security measure. Many a garden has been discovered by regular 12-hour peaks in energy consumption from power-hungry grow bulbs, or even the bulbs causing a fire. In contrast, LEDs, while expensive, last much longer than conventional grow bulbs, produce little heat, and may use less electricity than, for example, the peak consumption of a video game console.

Because indoor gardens with LEDs generally have stable, reasonable temperatures, plants undergo less stress and less water is needed to maintain a healthy soil moisture level. Some gardeners have found that replacing conventional lights with quality LEDs also increases bud potency.

There are a lot of shysters out there selling cheap LED lamps at rip-off prices, but Area 51 LED, Kind LED, and Advanced LED Lights, among others, are trustworthy companies.[66] I have no affiliation with these companies, outside of being a customer of Area 51 LED. These companies' products and customer service have many positive reviews on the Internet, and I have had great results with Area 51's LED grow light model SGS-160.[67]

For the electronically inclined, check out DIY LED light designs.[68]

Indoor Garden Checklist

Below is a list of the basic equipment needed for a discreet indoor garden:

- **Grow tent or converted closet**
  - The tent or closet must block out all light when the grow lights are off or growth will suffer
  - This includes blocking lights inside the grow tent such as power strip lights and other electronic lights
- **Intake fan with a CFM rating to replace the air in the grow space every 1 to 3 minutes**
  - CFM stands for cubic feet of air per minute. Multiply the width, length, and height of your grow space/tent to determine its cubic feet or volume. This number is used to select an appropriately rated intake fan (for example, a CFM rating of at least a third of the grow space's volume). Increase the calculated volume by 20% if using a carbon filter, because the fan will need to be slightly stronger because of the filter's resistance
- **Oscillating fan**
  - This keeps the air moving, for strengthening branches and avoiding mold
- **Carbon filter that matches the fan's CFM rating**
- **Combined digital thermometer and hygrometer**
  - The ideal temperature when the grow lights are on is between 70° and 75° F, and no lower than 60° F when the lights are off
  - Sustained humidity above 70% is a concern, but plants grown in living soils are mold resistant, and foliar application of AACTs allows good microorganisms to compete with mold for the same food
- **Grow light(s)**
- **Digital outlet timer for controlling the light cycle**

- **Small and large gardening pots/containers**
  - It's easier to water smaller containers for germination
  - Some gardeners start seeds out in solo cups with lots of holes in the bottoms and then transplant the young plants into bigger containers

Cannabis Seed Selection

Now onto the star of the show: cannabis.

A Cannabis plant's genetics is equally important as soil health. Starting with bad genetics will, at best, lead to mediocre results. Choose a seed company with a good reputation, such as Serious Seeds, Dinafem, Bodhi Seeds, or Mandala Seeds.[69]

If you cannot buy seeds locally, the next best option is having seeds shipped within the country that you live in to avoid meddlesome custom officers inspecting your seed package.

Cannabis has two main species: Sativa, which is associated with an uplifting, cerebral high, and Indica, which is valued for pain-relieving and other medicinal properties. These different effects are due to a strain's cannabinoid and terpene profile: Sativa strains are typically dominated by tetrahydrocannabinol (THC) and Indica strains have a mix of THC and cannabidiol (CBD). Some Indica strains contain mostly CBD (for example, Charlotte's Web), resulting in a medicinal plant that has little-to-none of the psychoactive effects associated with THC. Most strains have both Sativa and Indica genetics, providing a considerable variety between pure Sativa and Indica strains.

Two further variations are *feminized* seeds and *auto-flowers*.

Typically, only female plants grow from feminized seeds. Regular seeds are a mixture of male and female seeds, and the sex of the plant is typically not known until the early flowering stage, when the female plants being to bud (flower) and the males begin to grow pollen sacs. Male plants do not bud, but rather fertilize female plants, which results in buds with seeds. Fertilized females will divert some of their energy to seed production at the expense of bud growth. So once plants show male or female traits in the early flowering stage, male plants are removed from of the garden in order to prevent seeds in the flowers (of course, plant breeders want seeds, so they don't remove the males).

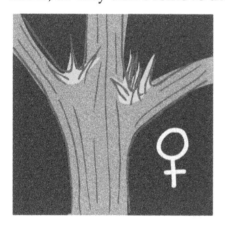 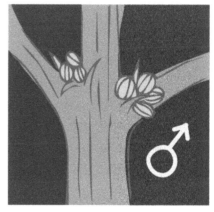

Auto-flowers are just that: Cannabis plants that flower automatically. Regular Cannabis plants typically do not enter the flowering stage until the light cycle is switched from 16 or more hours of light per day to 12 hours of light. But, auto-flowers do not require a particular light cycle to enter the flowering stage and will automatically produce flowers after a period of vegetative growth. Auto-flowers obtain this property by being a cross between Indica or Sativa and a third, less well known Cannabis species, Cannabis *ruderalis*, which provides the auto-flower genes.

There are even feminized auto-flowers, which would probably be one the most fool-proof and easiest ways to grow Cannabis for the first time.

There is a catch, particularly with auto-flowers. Seeds of a regular strain can be more robust and bigger yielders compared to the feminized or auto-flower counterparts, but there are plenty of stable and adequate-yielding feminized and auto-flower strains to satisfy most gardeners.

Growth Stages

After a seed is planted in a growth medium at a depth of about half of an inch (1.2 cm), a sprout typically appears within 5 days, especially when ambient temperatures are around 75° to 80° F. Not counting the week or two after seed germination, gardeners typically grow indoor Cannabis in the vegetative stage for 1 to 2 months. During this stage, the lights are kept on for a majority of the day. Light cycles range from 24 to 14 hours ON and 0 to 10 hours OFF, with 18/6 and 16/8 being the most popular settings.

To enter the bud-producing flowering stage, the light cycle is switched to 12 hours ON and 12 hours OFF (12/12), but some gardeners swear by 11 hours ON and 13 hours OFF (11/13). In either case, the shorter "days" trigger the plant's flowering stage.

The flowering stage typically lasts from 60 to 90 days after the light cycle is switched. Indica and Indica-dominated strains finish around the 60-day mark and Sativa strains typically finish around 70 to 90+ days.

As the flowering stage progresses, cannabinoid-rich trichomes spread and grow on the buds and leaves. Trichomes begin as clear, tiny offshoots and mature into a larger cloudy-to-amber resin that covers the buds as the flowering stage finishes.

Trichome development can be monitored with a magnifying glass, or with a digital camera by zooming into a captured image until individual trichomes can be seen.

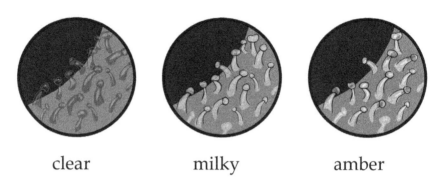

clear   milky   amber

To an extent, the timing of the harvest can also affect cannabinoid profile: in most strains, clear-to-milky trichomes will contain mostly THC, whereas CBD content will increase as the milky trichomes turn amber colored. Recreational gardeners tend to prefer around a 50/50 mix of milky-to-amber trichomes, whereas medicinal gardeners may prefer a mostly amber harvest for pain management. Of course, this balance is confined within a Cannabis plant's general cannabinoid profile, as discussed above.

Another sign of the late-flowering stage is when the leaves change color and fade from green to shades of red, yellow, and brown. This process is the visual indication that the plant is drawing nutrients away from the leaves for bud production, and it is a normal part of the process.

Gardeners using conventional salt-based fertilizers will "flush" a plant in the last 2 weeks before harvest by watering only with water, but flushing is not needed if you follow the natural gardening style of this book.

Harvesting

Once the trichomes have reached the desired color or mix of colors, the plant is chopped for harvesting. It is also possible to harvest the buds with the plant intact and regrow the plant by inducing the vegetative stage for a second round of bud production. I refer you to the Internet or a general guide for a more detailed discussion of this technique, known as *re-vegging*.

To harvest, cut the plant at the base and hang it upside down on, for example, a wire rack or wire music stand. The larger leaves can be pulled or cut off before or after this step. Next, working from top to base, cut off a section of the center stem connected to one or two side branches and trim the leaves jutting out from the bud. The connected stem and branches provide a V-shaped hook for hanging when drying. Small gardening shears work well for trimming. Make sure you scrape off the resin that builds up on the shears (scissor hash) for later use, such as putting in a vaporizer.

Once finished, you will have several V-shaped sections that can be individually hung, as shown below, or tied together in a sort of wreath for drying.

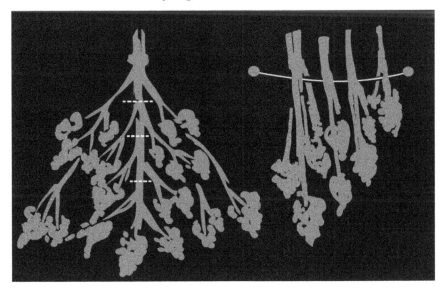

Drying

Depending on the drying room's temperature and humidity level, the buds will take 5 to 14 days to dry. A grow tent can also be a drying room, which is a useful approach for containing odors. Ideal drying room conditions include a temperature around 70° F, 50% relative humidity, and good air circulation.

After a few days, the buds can be lightly touched. Check for a slightly moist and sticky feel, which signifies that the buds are ready for curing. Curing further breaks down the remaining chlorophyll, which otherwise causes a harsh, unpleasant taste in fresh Cannabis. Curing also stabilizes the buds' moisture level, thereby readying the buds for long-term storage.

Another technique for determining how far along the buds are in the drying process involves cutting off the smaller, ancillary buds (popcorn buds) after 4 or 5 days, since they will dry faster than the bigger, denser buds. Place the popcorn buds into a glass jar with a small hygrometer (humidity meter), close the jar, and wait 8 to 24 hours. If the hygrometer registers 70% or higher relative humidity (RH), the drying process is not finished, and the jar should remain open for another day.

At or under 55% RH indicates that the other buds are ready to be placed in the jar, too. Although even 60% may be considered too dry, the bigger buds will share their extra moisture with the popcorn buds and even things out. The target curing range is between 55% and 65% RH, as measured after closing the jar for 24 hours. Once all the buds are jarred and are within this range, open the jar every few days for 30 minutes to an hour until the RH stabilizes around 55 percent and the jars no longer make a *pfft* sound when opened (that's why this process is called *burping*). After that, the jars can be stored and perhaps opened every few months.

Stealth

Ideally, buy all of your equipment locally with cash and do not send seeds to the same place you plan to garden.

The Internet is a great resource, but at least surf in private mode so that Google is not logging all of your searches. A virtual private network or proxy server service will also help protect against the automatic data aggregation that government agencies are wont to do. Also, pruned leaves can be placed in your worm farm or compost pile rather than thrown out in the garbage.

These small precautions are certainly not foolproof, but provide at least some protection from prying eyes. But, being stealthy doesn't mean being reclusive or shady. In fact, being a normal-to-affable guy with your neighbors goes a long way toward avoiding trouble.[70]

Eating (and Vaping) the Fruits of Your Efforts

"Vape" was chosen as Oxford Dictionaries 2014 Word of the Year, so odds are that you know what vaping is. In case you don't, vaporization heats Cannabis to a point that releases cannabinoids, but well below combustion temperatures. Not only is vaporization more efficient than combustion, it also releases considerably less carcinogens and odors. That said, while vaping is healthier than smoking, it's still not healthy in the traditional sense.

But, eating raw Cannabis buds certainly is,[71] even if doing this does not provide an obvious high. Butter or oil cooked with Cannabis, in contrast, can produce a strong, even trippy high since liver metabolizes THC into a more psychoactive form. If smell is a concern, use a pressure cooker for a discreet, low-smell cooking method.[72]

Speaking of eating, vaped material can have a second life as an edible. Save up at least a ¼ cup of vaped material, coarsely grind it, and soak it in olive oil for a month or two. Strain the oil and throw out the vaped material. The resulting oil, depending on the strength, can give a nice, mellow high from a tablespoon or two. Some people like to add vaped material along with fresh buds for making cannabutter or oil that provides both cerebral and corporal effects.

Note that fresh buds cannot be simply soaked in oil or high-proof alcohol like vaped leftovers can for producing a psychoactive oil or tincture. This is because the THCA (Tetrahydrocannabinolic Acid) present in fresh buds turns into psychoactive THC after being heated. So fresh buds, after grinding, need to be heated either directly in oil or placed in a covered Pyrex bowl and baked in an oven at 240° F for 60 minutes, preferably on top of a pizza stone to stabilize temperature fluctuations.[73]

For those wanting to know more about canna-cooking, The Stoner's Cookbook has a wealth of information on the subject.[74]

One last vapevantage is that you can vape your buds straight from the plant for testing purposes as harvest time nears. Let it sit, heated, for 30 to 60 seconds for the water content to evaporate. While it won't make for an ideal experience, vaping "green" bud is much preferred over smoking it.

The range of vaporizers have considerably grown in the last past years. High-end models range from units that fill balloons with vapor (Volcano and Herbalizer vaporizers) to all-glass vapor path models that cool the vapor with water and ice (VapeXhale Could Evo vaporizer).

Portable vapes include Ploom PAX, Arizer Solo and Air, and the Firefly. The Arizer Solo is the largest and least-sexiest of the bunch, but it can be connected to a bong via a glass adapter to cool the vapor with ice or water.

As you can tell, there is not a single "best" vape, but rather a plethora of choices to cater to your particular needs and preferences.

Endnotes:

66 http://kindledgrowlights.com/; http://a51led.com/; http://advancedledlights.com.

67 This model is no longer available, but is similar to the current model RW-150.

68 http://howtogrowmarijuana.com/diy-led-grow-light/

69 http://seriousseeds.com/; https://www.dinafem.org; http://www.aceseeds.org/; http://www.mandalaseeds.com/

70 http://www.growersunderground.com/attract-lots-unwanted-attention-garden.

71 http://www.thealternativedaily.com/weed-vegetable-juicing-best/

72 https://www.youtube.com/watch?v=vs66uyiH968

73 http://www.marijuanagrowershq.com/decarboxylating-cannabis-turning-thca-into-thc/

74 http://www.thestonerscookbook.com/

CHAPTER 6:
ORGANIC GARDENING IN EIGHT STEPS

Overview

Step 1. Make a soil mix.
Step 2. Moisten the soil mix.
Step 3. Let the soil mix incubate for at least 3 weeks.
Step 4. Plant a vegetable or Cannabis plant (or seed).
Step 5. Water the soil with an enzyme tea.
Step 6. Water the soil or spray leaves with a botanical tea.
Step 7. Once a month, inoculate the soil with an AACT.
Step 8. Recycle or amend the soil when plant growth diminishes.

As I have stated before, you can grow a kick-ass garden by just mixing (or buying) soil and regularly watering it. So, the eight steps listed above can be simplified to just four steps: mixing the soil (Step 1), letting the soil compost (Step 3), planting a seed or plant (Step 4), and watering. This can be further simplified if you buy a premixed soil that's similar to those sold by BuildASoil or KIS Organics.

If you're avoiding animal products in your garden, note that some commerical soil mixes will include fish bone meal, blood meal, and bat guano.

For DIYers, Step 1 (detailed below) provides a few alternatives for your soil mix, so don't feel like you have to include every ingredient listed under Step 1's *Soil Amendments*. A simple soil made from the base mix, neem, and glacial rock dust is still better than most commercial, bagged soils. Those ingredients, to me, are the essential components for nutrient-dense vegetable gardening.

Steps 2, 5, and 6, which cover compost, enzyme, and botanical teas, are additional tweaks to help your garden achieve its full genetic potential. If DIY botanical teas sound like a hassle, there are premade blends that you simply mix with water and apply. Examples include Kelp4Less' *Extreme Blend* and BuildASoil's *Craft Blend*. For those who just want to use bottled, liquid fertilizers, the Vegamatrix line by Pure Life Veganix appears to be designed to work with soil life, but it is not entirely organic and costs more than making your own teas.

That said, it's your garden so choose what works for you and ignore some of the holier-than-thou attitudes that come with organic gardening.

The last steps, monthly applications of AACTs (Step 7) and recycling or amending the soil (Step 8), are optional steps that may not even be needed, depending on the state of your soil. Well-established microbe populations living in ideal soil conditions won't need AACT supplementation as much as barren soils will.

If you're new to organic gardening, don't fret about doing everything "perfectly". There is no such thing. Plus, soil is a buffer and quite forgiving, so just don't be neglectful, and you'll be fine.

For example, I initially didn't have my watering technique down, and the top inch of soil in my 15 gallon containers was too dry for too long. Because of this, the compost microarthropods that I first saw crawling around my soil after each watering eventually vanished. My garden still turned out great. In fact, after harvesting,

I was pleasantly surprised, proud even, to see the compost worms (or their progeny) that I put in months ago, wiggling around in my garden pots, doing their thing.

So enjoy the process and learn from your mistakes. To paraphrase a wise cartoon dog, making mistakes is a natural consequence of trying to become good at something.[75]

The Eight Steps in Detail

Step 1: Combine your soil mix according to the following ratios.[76]

BASE MIX

* There are two options for the base mix
  — A ratio of 1:1:2 (one part compost, one part aeration element, and two parts peat moss)
  — A ratio of 1:1:1 (one part compost, one part aeration element, and one part peat moss)

Of the two ratio options, all equal parts (1:1:1) is great for a beginner gardener because the higher ratio of **aeration element** makes overwatering difficult to do, and the ratio option with more sphagnum peat moss (1:1:2) retains moisture for longer periods of time, so it works well for sprouting seeds

- Combine the following **Base Mix** ingredients according to one of the two options
  - One part **compost** (*most* important ingredient)
    - o Preferably, use "living"[77] earth worm castings or other high-quality compost
    - o Ideally, use a mixture of different compost types
    - o Low on compost? Try a combination of soil and compost, up to half soil (commercial or recycled) and half compost
  - One part pebbled-sized red lava rock, pumice, rice hulls, or other **aeration element**
  - One part or two parts **sphagnum peat moss**
    - o Hydrate before measuring and mixing, preferably using a wetting agent
- The total volume of the **Base Mix** in cubic feet dictates how much of each Soil Amendment you will add

SOIL AMENDMENTS

- Combine the **Soil Amendments**, then add to the **Base** Mix
- The given quantities are for each **cubic foot** of the **Base Mix** you made

- **Mineral mix** (*essential*)
  - 2 to 4 cups **rock dusts**, including at least one of the following: **glacial rock dust, basalt, bentonite, Azomite rock dust**, and/or **other minerals**
  - The **mineral mix** can include up to 4 cups **glacial rock dust**
  - No more than 1 cup each of **other rock dusts (basalt, etc.)** for each cubic foot of **Base Mix**
  - Mix and match depending on local availability

- **Meal mix** (*performance enhancers*)

A soil with 4 cups of meal mix typically only needs water from seed to harvest and may not need topdressings or botanical teas for the first crop. It also works well for small-container gardening (containers around 5 gallons)

2 to 4 cups total **meal mix,** containing the following
- 1/2 to 1 cup **neem meal** OR 1:1 mix of **neem meal** and **karanja seed meal**
  - o Improves plant immunity and soil balance
- 1/2 to 2 cups **kelp meal**
  - o Provides a broad spectrum of trace elements, among other benefits
- 1/2 to 1 cup **crab shell meal**
  - o Calcium source; pH buffer; high in chitin
- 1/2 cup **alfalfa meal** (*optional*)
  - o Source of nitrogen, among other goodies
- 1/2 to 1 cup **ground bugs or bug poop (frass)** (*optional*)
  - o Ground bugs and frass provide chitin
  - o Chitin's presence tricks a plant into strengthening its cell walls and excreting substances to ward off pests as a defense mechanism
  - o Only needed if you have a bad bug or soil pathogen problem
- 1/2 to 1 cup **all-purpose dry organic fertilizer** (*optional*)
  - o Not needed when you already have **kelp** and **crab shell**
  - o Example **fertilizers**: Epsoma Tomato-tone and Garden-tone, Happy Frog All-Purpose and Tomato and Vegetable, and similar products by Dr. Earth and Down to Earth

- **Calcium supplement** (*optional, but recommended for soils intended for multiple harvests*)
  - 1 cup **oyster shell flour** or **gypsum powder**
    - o Calcium source, for calcium-hungry plants; pH buffer
    - o Can use ½ cup of each for a total of 1 cup (*recommended*)
    - o Variable particle size will provide both immediate and long-term fertilization
  - Broken-up **oyster shells** can also be an **aeration element**

- **Biochar** (*optional*)
  - Add up to 10% of the **Base Mix's** volume
  - **Biochar** can be bought from gardening suppliers and should be broken up in small pieces
  - **Natural charcoal**, which is the variety made from hardwood (not briquettes), can also be used
  - Activate **Biochar** by mixing it with living compost or soaking it in compost tea for a few days before mixing with soil

Step 2: Moisten soil mix (wet, but not muddy) with nonchlorinated water.

- Instead of just water, moisten soil mix with an actively aerated vermicompost "tea" or other compost tea (AACT)
- To make compost tea, add for each gallon of water
  - 1/2 cup **worm castings** or other **living compost**
  - 1.5 tbsp **unsulfured blackstrap molasses**
  - Bubble for 36 hours with aquarium air pump and air stone

  When making more than 2 gallons of tea, an aquarium air pump may not be adequate

Step 3: Let the soil mix sit for at least 3 weeks on a tarp or in an open container.

- Preferably wait 4 weeks or longer
- Turn the soil once a week

TIP: The soil can compost or "cook" in the garden pot you will use to grow your plants in

TIP: After 2 weeks, plant a cover crop, such as microclover or Dutch white clover, in the soil. The soil should not be turned after planting a cover crop

TIP: After 2 weeks, throw some compost worms in the soil

Step 4: Place a vegetable or Cannabis plant (or seed) in a garden pot filled with the soil mix.

- After covering the plant's root ball with soil, add one or more types of mulches
  - Plant a cover crop such as Dutch white clover or microclover
  - Add a mulch of harvested (cut or mowed) **dynamic accumulators**
    - o **Dynamic accumulators** include dandelion, kelp, stinging nettle, comfrey, and watercress
    - o **Dynamic accumulators** may be on top of the soil or mixed in the top layer
  - Add a 1/2 to 1 inch topdressing of **worm castings**
- Thoroughly water the transplanted plant and soil

TIP: Use Air-Pots garden pots or fabric pots like GeoPot or Smart Pots containers

TIP: Sprinkle mycorrhizal fungi in the bottom of the soil hole before placing the plant's root ball inside the hole

TIP: Spray the plant's root ball with a kelp botanical tea (1/4 cup kelp per gallon) before placing the root ball in the soil; this helps prevent transplant shock

Step 5: Water the soil with an enzyme tea[78] at least once per week.

For each gallon of **enzyme tea**, perform the following:

1. Measure out 10 to 12 grams of **barley seeds, mung bean seeds, alfalfa seeds,** or other **quick-sprouting seeds.** When using alfalfa seeds, use only 6 grams and only during the vegetative stage of your vegetable or cannabis plant
2. Soak seeds in nonchlorinated water for 8 to 10 hours
3. Strain and rinse seeds. Get rid of the soak water because it contains growth inhibitors
4. Sprout seeds in a jar of aerated water, sprouting tray, or damp cloth
    a. Damp cloth method: soak a cloth with nonchlorinated water and lightly wring out; place seeds in cloth and cover by folding the cloth over the seeds; do not let cloth dry out
    b. After 3 to 5 days, the sprouts have grown the length of the seed, which is long enough
5. Place seed sprouts and some water in a blender and puree
6. Coarsely strain the puree, add 1 gallon of nonchlorinated water to the strained puree, and water plants with the resulting mixture

TIP: Enzyme teas can be alternated with plain water
TIP: Enzyme teas and botanical teas can be combined
TIP: Add 1 to 2 tbsp **BioAg Ful-Power fulvic acid** per gallon of enzyme tea. Use the **BioAg Ful-Power fulvic acid** only once per week

Step 6: Water the soil or spray leaves with a botanical tea no sooner than 3 weeks after sprout.

- Start out with once-a-week applications, but feel free to experiment with application frequency
- For each gallon of water,[79] add
  - 1 to 2 tbsp **alfalfa meal** (stop using alfalfa after 2 to 3 weeks into flowering), and/or
  - 1 to 2 tbsp **kelp meal**, and/or
  - 1 to 3 tsp **neem oil** (preferred) or 1 to 2 tbsp neem meal
    - o You can alternate applications of alfalfa, kelp, and neem
    - o All **three ingredients** may be combined
    - o Dilute a combined tea for young or sensitive plants
    - o Choose only **one ingredient** for foliar sprays, as leaves are particularly sensitive
- Soak or bubble the above ingredients in water for 24 to 72 hours with an aquarium air pump and air stone
  - o **Neem oil** does not need to be soaked and should be added right before application
- Strain the meal(s) from the botanical tea and apply the botanical tea to the soil or as a foliar spray

TIP: Place the meal(s) in a paper tea filter to avoid clogging sprayers and watering cans
TIP: The strained leftovers can be put in a worm bin

- **Enzyme teas** and botanical teas can be combined
- **Botanical teas** can be alternated with plain water for watering the soil

BOTANICAL TEA ADD-ONS

- For each gallon of **botanical tea**, you can add the following:
  - A commercial **mycorrhiza** product according to product's instructions
  - 1 to 4 tbsp of **pure aloe vera gel**, OR
    1/4 tsp **aloe 200x concentrate powder**, OR
    1/4 cup **aloe vera juice**, OR
    1/16 tsp **yucca extract**, OR
    a drop of **mild liquid soap**
    - o Mix well in **botanical tea** right before application
    - o **Aloe vera** and **yucca** are wetting agents, which are needed for foliar sprays
    - o **Pure aloe vera gel** can be store-bought or obtained from an aloe vera plant
    - o **Soap** is needed only for foliar sprays if you don't have **aloe vera** or **yucca**
      - **Castile soaps** work well with this application
  - 1 tsp **Dyna-Gro Pro-TeKt liquid supplement**
    - o **Dyna-Gro Pro-TeKt** liquid supplement has silica, which is great for plants
      - **Agsil 16H silica powder** is a cheaper alternative
  - 1/4 cup pure **coconut water**
    - o ½ cup of **coconut water** can be used for a plant's flowering stage
    - o **Coconut water** can be added to plain water for soil applications

- 1 to 2 tbsp **BioAg Ful-Power fulvic acid** OR
  1/4 tsp **BioAg TM-7 humic acid**
    - o   Either can be added to plain water for soil applications, but only one soil application per week of either acid
        - When using both products for soil applications, alternate weekly between each product
    - — **BioAg Ful-Power fulvic acid** can also be used in foliar sprays as it is liquid, so no soaking/ bubbling is needed
    - — **TM-7 humic acid** can be bubbled with kelp, alfalfa, or neem meals

Step 7: Once a month, reinoculate the soil with an AACT (optional; see step 2).

- Compost teas may not be needed if the soil still has an abundance of microbe soil life from living under ideal conditions
- In contrast, weekly AACT applications are not a bad idea when you are unable to access high-quality compost

Step 8: Recycle or amend the soil when plants grow less robustly (optional).

- Break up the old root ball and mix it in with old soil when recycling the soil
- **Light recycling**[80]
  — For each cubic foot of used soil, add
    o 1 part **compost** for every 4 parts old soil
    o half the amount of **aeration element** compared to the amount of added compost
    o 1/2 to 1 cup **kelp meal**
    o 1/4 to 1/2 cup of **all-purpose dry organic fertilizer**
    o 1/4 cup **oyster shell flour** or **gypsum powder**
    This light amendment can be immediately used for your next crop
- **Heavy recycling**
  — Add compost and aeration element just like in the Light Amendment option
  — Add any of the ingredients listed under Step 1
  — Repeat Steps 2 through 6
- **Easy recycling**
  — Add the old soil as a portion of the **Base Mix's** compost component of Step 1, up to a 1:1 ratio of compost and old
  — Repeat Steps 2 through 6
- **No recycling**
  — Topdress the soil with the following:
    o Just **vermicompost** AND/OR
    o 1/2 cup **mineral mix** and 6 tbsp **meal mix**
    Spread on top of the multi-harvest soil or mix with compost or other soil beforehand and then spread on top of the multi-harvest soil

Endnotes:

75 Jake the Dog, from the *Adventure Time* episode "His Hero". (In this exchange, Finn is lamenting his failed attempts at solving problems nonviolently when Jake beautifully casts failure in a positive light: **Finn:** Apparently, I suck at being nonviolent. **Jake:** Dude, suckin' at somethin' is the first step towards bein' sorta good at somethin'. You and I are like little baby Billys right now, and we're "sucking" on our first bottle of nonviolent milk!)

76 Credit goes to Headtreep, Cann, Ganja Girl, Greasemonkeymann, and ClackamasCootz, of the Recycled Organic Living Soil Internet boards, for the particular ratios and combination of ingredients.

77 "Living" means moist compost with active bacterial and fungal populations; completely dry is useless.

78 Credit goes to ClackamasCootz for adapting this technique from beer brewers.

79 To convert to liter of water, use the same number of teaspoons per liter (e.g., 1 to 2 teaspoons per liter instead of 1 to 2 tablespoons per gallon).

80 Credit goes to Ganja Girl for the particular ratios and combination of ingredients for soil recycling.

AFTERWORD

Thank you for reading my book. I truly appreciate it, and I believe you have saved many hours of research which I conducted to get to the same place. However, do not take this book as gospel. I believe my advice is sound, but advice does not replace experimentation and further education. Also, plant preferences are as varied as human preferences, and only a side-by-side comparison can definitively prove whether one technique is superior to another for a particular plant.

Thus, I've tried to share the various options for you to try, improve, or forgo. Perhaps you will get great results using only water and enzymes teas with a buried top layer of dandelion leaves.

I am indebted to the many participants of the Recycled Organic Living Soil Internet boards. These organic alchemists taught me how to garden with Nature instead of replacing Nature with commercial products. Without their online contributions, this book would not be possible.

Feel free to contact me at **zonderkop@hush.com.** I cannot promise I'll respond, but I will appreciate the feedback.

Happy gardening.

—Kip Zonderkop

FURTHER READING
AND RESOURCES

References:

- *Teaming with Microbes: The Organic Gardener's Guide to the Soil Food Web* by Jeff Lowenfels and Wayne Lewis
- *Teaming with Nutrients: The Organic Gardener's Guide to Optimizing Plant Nutrition* by Jeff Lowenfels
- *The One-Straw Revolution: An Introduction to Natural Farming* by Masanobu Fukuoka
- *The Best Place for Garbage: The Essential Guide to Recycling with Composting Worms* by Sandra Wiese
- *The Intelligent Gardener: Growing Nutrient Dense Food* by Steve Solomon
- *Earth Repair: A Grassroots Guide to Healing Toxic and Damaged Landscapes* by Leila Darwish
- *Bioshelter Market Garden: A Permaculture Farm* by Darrell Frey
- *Nourishing Traditions: The Cookbook that Challenges Politically Correct Nutrition and the Diet Dictocrats* by Sally Fallon and Mary Enig

General Internet Resources

Key search terms: "Recycled Organic Living Soil" or "ROLS", "no-till gardening", "permaculture".

Selected Links Index

Cannabis Juicing

http://www.thealternativedaily.com/weed-vegetable-juicing-best/

CannaCooking

http://www.thestonerscookbook.com/

Carbon Filters

http://greengatorfilters.com/

Chitin

http://www.onfrass.com/

Compost Tea Brewing Systems

http://www.kisorganics.com/

http://microbeorganics.com/

Decarboxylating Cannabis

http://www.marijuanagrowershq.com/decarboxylating-cannabis-turning-thca-into-thc/

Dynamic Accumulators

http://soulflowerfarm.blogspot.com/2012/11/dynamic-accumulators.html

http://en.wikibooks.org/wiki/Permaculture_Design/
Soil#Dynamic_Accumulators_of_Nutrients_for_Composting

http://buildasoil.com/blogs/news/9745744-diy-nutrients-from-dynamic-accumulator-plants

LED Grow Lights

http://area51lighting.com/

http://kindledgrowlights.com/

http://www.apachetechinc.com/

http://advancedledlights.com/

http://californialightworks.com/

Microarthropods

http://www.pacifichorticulture.org/articles/soil-microarthropods/

Organic Gardening Education

http://theearthproject.org/

http://rodaleinstitute.org/

http://edibleschoolyard.org/

Pit Greenhouses

http://www.the-meal.net/graph/manuel_walipina_benson.pdf

http://www.undergroundhousing.com/greenhouse_book.html

Pressure-Cooker Cannabis Oil

https://www.youtube.com/watch?v=vs66uyiH968

Raised-Bed Garden

http://www.wikihow.com/Practice-Square-Foot-Gardening

Seeds

Bodhi Seeds

http://provisionseeds.com/

http://www.mandalaseeds.com/

http://www.aceseeds.org/

http://seriousseeds.com/

https://www.dinafem.org (ships from Spain)

http://www.naturesgreenremedies.com/ (ships from the USA)

http://shop.ilovegrowingmarijuana.com/ (ships from the Netherlands)

http://www.peakseedsbc.com/index.htm (ships from Canada)

http://www.sensibleseeds.com/ (ships from the UK)

https://www.seedbay.com/ (cannabis seed auction site)

Soil Ingredient Sources

Local feed, landscaping, and gardening stores.

eBay and Craigslist.

http://buildasoil.com/

http://www.kisorganics.com/

https://www.kelp4less.com/

http://www.planetnatural.com/

Soil-Moisture Meter

Bluemat Soil-Moisture Sensor

Teas

http://www.frenchgardening.com/tech. html?pid=3164873867231346

http://buildasoil.com/blogs/news/12607517-using-b-a-s- barley-for-enzyme-tea-tutorial

http://theunconventionalfarmer.com/

http://www.kisorganics.com/

http://microbeorganics.com/ (excellent resource on compost teas)

https://www.kelp4less.com/

Technical Assistance and Awareness

http://www.2600.com/

Vaporizers

http://www.volcanovaporizer.com/

http://www.herbalizer.com/

http://www.vapexhale.com/

https://www.ploom.com/

http://www.thefirefly.com/

http://arizer.com/